HOME WI-FI TUNEUP

Practical Steps You Can Take to Speed Up, Stabilize, and Secure Your Home Wi-Fi

A Short Read By

MARLON BUCHANAN

ISBN: 978-1-7355430-3-1 (paperback)

ISBN: 978-1-7355430-2-4 (ebook)

HomeTechHacker.com

Other books by Marlon Buchanan:

The Smart Home Manual: How to Automate Your Home to Keep Your Family Entertained, Comfortable, and Safe

CONTENTS

PREFACE

Wi-Fi in our homes is no longer a luxury, it's a necessity. Unfortunately, there are few things more frustrating than bad Wi-Fi connections. Who wants their streaming video to keep pausing and buffering every 5 minutes? Or to randomly drop out of Zoom calls?

You don't have to suffer from bad Wi-Fi in your home. There is a fix for almost every Wi-Fi problem and this guide aims to give you the tools and techniques you need to diagnose and fix your Wi-Fi. Just follow the steps in this book and you'll be much happier with your home Wi-Fi.

BONUS MATERIAL

This is a concise, practical guide to help you with your Wi-Fi. If you are looking for more in-depth information about Wi-Fi, home networking, smart homes, and other home technology topics be sure to subscribe to HomeTechHacker.com. I regularly publish how-tos and reviews that will interest you.

INTRODUCTION

Have you noticed how important the Wi-Fi connection in your home has become? With the rise of mobile devices and the Internet of Things (IOT), your home Wi-Fi is more important than ever. Cell phones, streaming devices, smart bulbs, home security systems, washers and dryers, smart speakers, TVs, and so many other gadgets now require Wi-Fi connections to access all the geeked out features that caused you to buy them in the first place.

The significant reliance on your Wi-Fi means it's important to make it fast, secure, and stable. This book will discuss the three major problems most people have with their home Wi-Fi (some of which you may not even know you have) and more importantly, detail ways to diagnose and fix these problems.

Before we start, let's define a few terms so we are all on the same page:

- **IoT – Internet of Things.** A collection of interconnected and interrelated devices that can communicate and transfer data over various networks without human interaction. It most commonly refers to the explosion of devices that are everywhere that connect to the Internet. Examples include Smart TVs, smart speakers, toys, wearables, smart appliances, and smart meters.
- **Wireless Access point.** A wireless access point (sometimes just called an access point or an AP) connects to a router, switch, or hub via Ethernet, and projects a wireless signal to a designated area. These are used for extending a wireless network.
- **Router**. A router forwards (or routes) packets of data between different networks. For instance, the router in your home routes packets coming from the Internet to your private home network and vice versa.
- **QOS – Quality of Service**. This refers to settings on your router that allow for the control of priority and treatment of different types of traffic (streaming, web browsing, torrenting) on your network.

MAJOR PROBLEM #1: YOUR WI-FI IS TOO SLOW

Is your Wi-Fi slow? Do you sometimes find yourself waiting a long time for your app to load, or an attachment to download, or for a web page to come up on your mobile phone? Do your favorite streamed shows buffer and/or stream at lower than HD quality? Let's make your Wi-Fi speed problems a thing of the past!

The root cause of your Wi-Fi slowdown can be hard to find. We are going to start with how to diagnose the major culprits and then apply a fix.

Wi-Fi Speed Culprit #1: Your Internet connection is slow

Symptoms:

1. You don't pay much for your Internet service.
2. Your Internet connection is really slow when multiple people are using it.
3. You haven't properly configured your router's QOS settings.

Diagnosis:

1. First, find out what bandwidth you have. Go to speedtest.net from a wired computer on your home network and test your Internet speed a few times at different points in the day. Make sure no other devices are doing any significant downloading or uploading to the Internet (streaming videos, downloading large files, etc.) at the same time. Are you getting the speeds you expect based on what you purchased from your Internet service provider (ISP)?
2. What speed did you pay for? Do you know how much speed you need? For example, Netflix recommends 5 Mbps for each high-quality HD stream and 25 Mbps for each Ultra HD (4K) stream. Below are some common bandwidth needs for home network activities:
    - HD video streaming: 5-25 Mbps per stream (downloading)
    - Music streaming: < 1 Mbps per song

(downloading)

- Online gaming: 3-6 Mbps download/1-3 Mbps upload per person
- Web/Social media: 3 Mbps download/1.5 Mbps upload per person
- HD video call (e.g. Zoom): 1-4 Mbps upload/download per person

3. The amount of bandwidth your home needs depends on the number of people and their regular usage of the Internet at home. For example, let's say it's common in your home for a couple of people to be streaming a video (~15 Mbps), a couple more to be browsing the web/ social media (~6 Mbps) while someone is gaming online (~6 Mbps) and streaming music in the background (~1 Mbps). 15 + 6 + 6 + 1 adds up to 28 Mbps which would be the minimum download bandwidth needed to support these activities. You should always add at least 30% overhead to your calculated minimum because the speeds ISPs quote are max figures and not what they consistently give you. Also, often there are other devices on the network that are steadily using small amounts of Internet bandwidth (e.g. checking for and downloading updates).

   Keep in mind that if you are frequently viewing and/or sending videos, pictures, and other files via social media your needs may be higher than the average 3 Mbps figure above. Also, there are often spikes in the rates needed by these services. For example, an online game may average 3 Mbps but may have certain spikes where it needs 10 Mbps to function properly. In the example I gave earlier I'd recommend you get around 40 Mbps to cover your needs.

If you are constantly downloading content (via torrents and/or Usenet, for example) then you already know you have high bandwidth needs. It will come down to how fast you want those downloads to happen and how often you download.

4. Have you looked at your router's quality of service (QOS) settings? These settings can determine which types of Internet traffic and which devices have priority so that you can use your Internet bandwidth efficiently. For instance, you might want to prioritize streaming video and video chats over torrent downloads and web surfing because you would like to keep the quality of video high and eliminate buffering pauses at the expense of slowing down a web page load or file download. These are the types of scenarios you can configure with your router's QOS settings.

After walking through the steps to test your Internet speed and estimate your needs do you find you have a fast enough Internet connection? If not, this could be a major reason your Wi-Fi seems slow. You must have enough bandwidth for all the devices (not just wireless) that access the Internet. If you don't, it's time to upgrade your Internet connection. You also want to make sure you have modified your router's QOS settings to prioritize the Internet traffic that is most important to you.

Wi-Fi Speed Culprit #2: Other devices are hogging your Wi-Fi bandwidth

Symptoms:

1. There is a lot of Wi-Fi traffic going across your LAN (Local Area Network).
2. Somebody is on your Wi-Fi network you don't know about...

Diagnosis:

1. Do you have a NAS (Network Attached Storage)? If you do, chances are you have backups going across your network. You are also probably streaming music, pictures, and videos wirelessly on your network using something like Kodi or Plex. These will hog a lot of bandwidth on your Wi-Fi.
2. Do you run some type of networked home surveillance system or do you have some wireless cameras in your home? Having these on your network can take up more bandwidth than streaming Netflix.

3. Have you run security checks and malware scans on the devices on your network? If not, you could have malware that is sniffing across your network doing nefarious things. In addition to compromising your devices and data, this can slow down your network!
4. Can you identify every device that is connected to your network? You want to make sure you don't have your neighbors connecting to your network and using up your bandwidth.

Think about the diagnosis items. Make sure you know what is going on in your network and what the impacts may be. If you have items on your network constantly sending traffic (like video cameras) try to make them wired devices instead of wireless. Wireless may be convenient but if there is an option for a wired connection you should take it and reserve the wireless bandwidth for mobile devices and others that are hard to give wired connections.

Go to your router and/or access point settings and see if you can tell which devices are connected to your wireless network and what bandwidth they are using. Many routers and access points have traffic charts to detail network activity. I'll talk more about detecting and preventing malware and unwanted users on your network in the security section of this guide.

Wi-Fi Speed Culprit #3: Some of your wireless devices are old

Symptoms:

1. You still have 802.11g devices on your network.
2. Your devices don't support the 5 GHz protocols.

Diagnosis:

1. What protocols do your mobile devices support? If you have a lot of devices that only support 2.4 GHz protocols (802.11g and 802.11n 2.4 GHz) your speeds are limited, and your Wi-Fi is subject to more interference than the 5 GHz protocols. Moreover, you may be forced into running your wireless router/AP in a mode that lowers the Wi-Fi speed of all devices that connect to it just to support the older devices.
2. What protocols does your wireless router/AP support? Does it have multiple radios capable of separating the protocols so that it doesn't need to run in a "mixed-mode" that slows down speed for the faster devices?
3. How long has it been since you upgraded your router/AP? Your router may be too old to support the newer protocols your mobile devices are capable of using. The age of

your router/AP also has security implications (more on that later).

If you have a lot of older devices on your network and you have to support older wireless protocols you are probably slowing down your network speeds. You might consider upgrading your router/AP to one that can support older and newer protocols without slowing down the faster devices and/or upgrading your older Wi-Fi devices. Check out HomeTechHacker.com's Buyer’s Guide for new router recommendations. Make sure to take the wireless protocol support of a device into consideration when making a purchasing decision, both for your router and your client devices.

Wi-Fi Speed Culprit #4: Your Wi-Fi signal is weak

This is a common cause of slow Wi-Fi speeds. We will talk more about signal strength and stability later in this book.

MAJOR PROBLEM #2: YOUR WI-FI SECURITY IS LACKING

One of the easiest ways to hack into a home network is to attack through its Wi-Fi network. Have you taken steps to secure your Wi-Fi? Many pages of text could be devoted to ways to secure your Wi-Fi. In this book, I'll focus on actionable and important ones that will keep you covered against most vulnerabilities.

Wi-Fi Security Culprit #1: Your router is too old

Symptoms:

1. New firmware for your router hasn't been released in years.
2. Your router doesn't support dual-band wireless.
3. You don't get the Internet speeds you think you should get from your ISP.
4. You have to reboot your router frequently to keep it working properly.

Diagnosis:

1. Is your router more than 5 years old? Check to see what the latest firmware release is for your router. If it has been a few years, your router may not be able to provide adequate and up-to-date security.
2. Does your router support 5 GHz protocols like wireless N and AC? If it doesn't it may be too old.
3. Did you run the speed tests mentioned earlier? Were your speeds significantly lower than the package you bought from your ISP? There could be a few reasons for this, but one reason might be that your router is too old to provide the speed that you are paying for.
4. Do you have to reboot your router regularly for it to keep functioning?

If your router is running on old software it won't have the latest security protections. Make sure and check for newer firmware for your router.

If your router is outdated consider upgrading your router to properly protect your network and possibly give you faster Wi-Fi speeds and more capabilities.

Wi-Fi Security Culprit #2: You haven't changed the default settings on your router

Symptoms:

1. Your router login username and password are something like "admin" and "password."
2. The name of your Wi-Fi network (SSID) includes the brand of the router.
3. Your router's firewall isn't enabled.
4. You are not using WPA2 or WPA3 security.
5. Your router admin interface is accessible on the Internet.

Diagnosis:

1. Log in to your router (is the username/password "admin"/"password"?). Check to make sure that your firewall is enabled, you are using WPA2 and/or WPA3 security on your Wi-Fi, and that you have a non-default name for your Wi-Fi network. Find the setting to disable remote access (access from the Internet) to your router.

There are many more steps to properly secure

your router, but if you take care of these your Wi-Fi network will be much more secure. Leaving settings at the defaults allows hackers to more easily compromise your home network. Many hackers target specific brands of routers and leaving the brand in your Wi-Fi name signals to hackers the brand of your router as well as that it may have other default settings they can exploit.

Wi-Fi Security Culprit #3: You don't have a guest network

Symptoms and diagnosis:

The symptoms and diagnosis here are simple. You never set up a guest Wi-Fi network and/or your router/AP doesn't have that functionality.

If you don't have a guest network, you may find yourself in a situation where you have guests over and you have to give them your password so they can access your network. This gives your guests not only your password, which may be sensitive and used for different accounts, but also access to all of your networked devices (printers, network shares, PCs, etc.). Sure, you probably trust the people who have your password, but do you completely trust that all the software and apps they use are free of malware that can compromise the devices on your network? You shouldn't. A proper guest network allows you to set a separate and shareable password for your guests and firewalls off your private computer resources from being compromised by guests. Many routers and APs have this feature. If yours does it's worth learning how to configure it. If not, consider purchasing a router or AP that has the feature.

Wi-Fi Security Culprit #4: You have not segregated your IoT devices from the rest of your network

Symptoms and diagnosis:

The symptoms and diagnosis are simple here as well. If your IoT devices are running on the same Wi-Fi network as all of your other devices they probably have access to the rest of your network. Unless you've done something explicitly to limit their access (VLANs, access control lists, a separate IoT Wi-Fi network) they are probably members of your network like any other device.

You may be asking why IoT devices need to be segregated from the rest of the network. The answer is that IoT devices have been hacked en masse. Lightbulbs, switches, thermostats, refrigerators, network cameras, and many more IoT devices have been hacked. Some have been compromised to attack other people and companies, some have been used to attack the device owners. IoT devices have revealed owner Wi-Fi passwords.

IoT devices are not known for having the best

security features, and they are usually always connected and accessible from the Internet. Once these devices are compromised they can be used to probe and attack other devices on your network. Sensitive personal information stored on your computer could be compromised. This is why you want to protect the rest of your network from these devices.

Segregating your IoT devices can be a complicated affair. Some ways of doing it involve modifying routing tables and setting up separate subnets. But nowadays newer routers and access points have features where you can set up a network just for these types of devices that secure them from your other network devices. This works very similarly to a guest network. You just end up putting your IoT devices on a separate network. When you are looking for a new router or AP make sure and look for this function. Yes–we live in a world where we need to protect ourselves from lightbulbs and thermostats!

I have spent a lot of time segregating my network because I have a lot of smart devices. I used ACLs and firewall rules to accomplish my goal. Mine is a bit more complicated because there are sets of devices I want to be able to access the IoT devices and I want the IoT devices to have access to them, but not all non–IoT devices. As I said, it's complicated. Here are some more tips to improve your home network security.

MAJOR PROBLEM #3: YOUR WI-FI SIGNAL IS NOT STABLE THROUGHOUT YOUR HOME

Wi-Fi Stability Culprit #1: Your router is your wireless access point

Symptoms and diagnosis:

Most routers also have wireless AP capabilities. This can be a cost-effective way, especially for smaller homes that only need one wireless AP, to provide coverage for the entire house. However, the lack of flexibility with where you can place the router can be a problem. For most homes, the ideal place to put a single wireless AP is somewhere near the middle of your home high up on a wall or the ceiling. This spot is usually not an option for most homes because routers need to be near the modem provided by the ISP, which is often in a room nowhere near the middle of the house. Also, many people have a modem provided by the ISP that also functions as the router and AP, further constraining its placement.

If you can't place the router in a good spot in your home, or you have a larger home that a single router can't cover, you will have spots in your home that aren't covered or have poor wireless performance. You can fix this by buying a Wi-Fi range extender, a wireless access point, or by replacing your router with a mesh system. Let's take a quick look at these options.

A range extender, such as the TP-Link AC1750 Wi-Fi Range Extender, simply rebroadcasts your wireless signal to other parts of your home, thereby extending your wireless coverage range. Some have extra features like providing an ethernet connection and providing a guest network. The downside of range extenders is that there will be some speed loss, but some wireless extenders greatly reduce this loss by using multiple radios (separating the duties of receiving and transmitting the signal) and faster processors. Look for a range extender with these features.

A wireless access point, like the TP-Link EAP 225 (see my review here), plugs into a wired port somewhere in your home and can broadcast a wireless signal from there. Unlike range extenders, there is no wireless speed loss because they aren't using the Wi-Fi signal to also transfer data to your router. The downside, when compared to range extenders, is that they need access to an ethernet connection to work. Many also have features like guest networks and traffic monitoring. If you have access to a wired connection in an ideal spot in your home this is my recommended option. If you have a really large home you can use multiple access points.

Mesh Wi-Fi systems, like the Netgear Orbi or Google Wifi, have become popular. You can think of them as multiple easy to setup range extenders that are made to work together. Most systems give advice and tools on how to place them to get full coverage for your home. One key piece of information to know is that mesh systems are also designed to replace your router. This means they come with tons of features that range extenders will not have. If you don't want to or can't replace your router then mesh systems may not be for you (unless you want to get into the complications of running multiple routers on your network, or you get a mesh system that can work in bridge mode). This is the most expensive of the three options. If you have a large home that doesn't have wired networking throughout the house this is probably your best option.

If you are looking into buying a range extender, access point, or mesh system be sure to check out my Buyer's Guide for recommendations.

Wi-Fi Stability Culprit #2: You have too many Wi-Fi devices on your network

Symptoms and diagnosis:

Diagnosing whether or not you have too many Wi-Fi devices on your network can be difficult. The issue I am diagnosing here is different than too many devices hogging your LAN or Internet bandwidth that I wrote about earlier. Theoretically, most wireless access points can have 255 devices connected at a time, but that doesn't mean they'll work well with that many. You probably won't find the maximum limit in your access point's documentation. If you only have one access point I'd start to worry about having too many Wi-Fi devices when you get to around 20-25 devices connected at a time. At this point, depending on how active those devices are, you may also start to run into bandwidth issues. I mentioned how to solve the bandwidth issue earlier, but if you believe you have fixed your bandwidth issues and you are still experiencing slowness you may want to upgrade your access point or add an access point to distribute the load of your wireless devices.

Some of you are thinking that you'll never get to 20-25 wireless devices. It can happen faster than you think. Smartphones, tablets, streaming sticks, video game consoles, smart TVs, smart hubs, smart bulbs, thermostats, smart plugs, smart refrigerators, smart toothbrushes, smart scales, smartwatches, smart smoke detectors, smart washers, and dryers... I think you get my point. More and more Wi-Fi enabled devices are entering your home.

Wi-Fi Stability Culprit #3: You aren't using the best Wi-Fi protocols

Symptoms:

1. You only have 2.4 GHz wireless.
2. You don't have wireless AC.
3. You have Wi-Fi 6, wireless AC, or 5 GHz wireless N but your client devices don't use them.

Diagnosis:

1. Go check the wireless settings on your router. Do you have 5 GHz N available? Do you have wireless AC available? Wi-Fi 6? Are any of your devices connecting to the 5 GHz protocols?

2.4 GHz Wi-Fi (both G and N) can be very unstable in today's world. Those protocols deal with interference from other devices (like microwaves, phones, security systems, and baby monitors) and neighbors running 2.4 GHz wireless networks. This interference can cause speed slowdowns and even drops in your Wi-Fi connection. The 5 GHz protocols are much more resistant to interference. They are also faster. The downside is that they don't cover as much area as the 2.4 GHz protocols, and fewer clients support them. My suggestion would be to avoid buying devices that don't support 5 GHz protocols if possible from here going forward.

FINAL WORDS

Stable and fast Wi-Fi is a critical component of your home technology infrastructure. Few things are more frustrating than constantly buffering Netflix streams and slow mobile browsing. An insecure home network can lead to disastrous consequences. Hopefully, I've provided you with tips on how to diagnose and improve your Wi-Fi speed, security, and stability.

HOME WI-FI TUNEUP CHECKLISTS

Wi-Fi **Speed and Stability:**

- ❑ You know your Internet bandwidth needs
- ❑ You have enough Internet bandwidth
- ❑ You are using wireless AC 5 GHz and/or Wi-Fi 6
- ❑ You have a strong Wi-Fi signal everywhere you need it
- ❑ You know all the devices on your Wi-Fi network
- ❑ You know what devices are using a lot of bandwidth on your Wi-Fi network
- ❑ Your router supports the number of wireless devices you have on your network
- ❑ You have configured your router's QoS settings

Wi-Fi **Security:**

- ❑ Your router firewall is turned on
- ❑ Your router is less than 5 years old or still gets frequent updates
- ❑ Your router firmware is up to date

- ❑ You are using WPA2 or WPA3 security with AES
- ❑ You have a strong Wi-Fi password
- ❑ You've changed your Wi-Fi password in the last year
- ❑ You have a guest network
- ❑ You've isolated your Internet Of Things (IoT) devices
- ❑ Your router web interface isn't accessible from the Internet

ABOUT THE AUTHOR

Marlon Buchanan is an IT Director by day and a writer and home technology hacker (the good kind) by night. He has over 25 years of combined experience as an IT director, software developer, systems analyst, college instructor, and smart home consultant. He has a bachelor's in computer science and engineering and master's degrees in business administration and software engineering.

He has been automating things around his house since he was a kid. Now, his wife and kids get to enjoy the fruits of his smart home exploits. He is best known for his smart home, home networking, and cord-cutting articles on his blog HomeTechHacker.com and his book, The Smart Home Manual.

FEEDBACK AND NEXT STEPS

I hope you learned a lot from this guide and that you've improved your Wi-Fi. Also, I'd love to hear from you! Please take a couple of minutes to leave a review of this book on Amazon. Your review will help others know whether or not this book is for them. It will also provide me with information that will help improve my future books.

Also, if you are into smart homes, please check out my book The Smart Home Manual. It's a comprehensive guide to help anyone turn their home into a smart home.

Further info: HomeTechHacker.com
Twitter: @HomeTechHacker

Made in the USA
Monee, IL
06 November 2026

68010598R00023